F OR GOOD REASO, who have taken ar found a professional's booked a place to sta, through Airbnb love the sharing economy. These services, and countless others like eBay, TaskRabbit, Rent the Runway, and GoFundMe have a common theme – they connect people though online platforms. Often, the sharing economy enables transactions that used to be infeasible, time consuming, or cost prohibitive.

Others see the sharing economy as an enemy to be eliminated. It is true that the work done in the sharing economy does not look much like that of 1950s-era taxi drivers, artisans, repairmen, or hotel managers, but a twenty-first century economy cannot flourish under a decades-old policy framework. As technology continues to disrupt established business models, fights between innovators and special interests will only intensify.

Opponents of the sharing economy do not want to promote innovation; they want to protect themselves from change and competition.

And political opponents refuse to understand that the fundamentally different businesses of the new economy require a modern regulatory model to thrive and fulfill their potentials. Instead of embracing these advances, these groups resist any change that threatens their short-term economic interests.

To fight back, it is critical to illuminate the new economy's benefits and reveal the motives and tactics of those who desperately want to kill it. Change is central to economic growth, and Americans should not remain quiet while policy makers limit innovation to protect the politically powerful.

The Twenty-First Century Economy

Before highlighting political opposition to the sharing economy, it is important to understand the recent economic developments that are behind the business models of companies like Uber and Airbnb.

Underpinning these changes is the economic concept of "transaction costs." Nobel

Prize-winning economist Ronald Coase explains that transaction costs are what it takes to bring together buyers and sellers, exchange information, negotiate prices, and enforce contracts.

In the past, transactions were heavily influenced by businesses. But all that has changed in today's economy because technology has put information in the hands of consumers.

As the Competitive Enterprise Institute's Iain Murray argues, the sharing economy is made possible by the substantial lowering of transaction costs resulting from recent technological advances. Peer-to-peer interaction over the Internet gives consumers and providers of goods and services the ability to quickly connect with and gather information about others.

Across all sectors of the economy, technology creates entrepreneurial opportunities for anyone with productive resources to offer. These resources can be anything from physical or intellectual services (such as handyman jobs, academic tutoring, or legal advice) to the use of property (be it a drill, car, or spare room).

On one hand, the needs served by sharing-economy companies are nothing new. There have always been people who wanted to buy a hard-to-find product, find a place to stay, access cash, eat a home-cooked meal, find assistance on a task, or figure out a way to get around. The problem was finding someone who was willing to offer the desired goods or services at a reasonable price. Imagine what it would have been like if people went from door to door asking home owners if they had an extra room to rent and for how much. Now, travelers can simply log on to Airbnb and, with a few clicks, find a room that fits their needs and budgets. Similarly, hitchhiking – a primitive version of ride sharing – was never a very efficient way to find rides.

The expansion in commercially available goods comes from the ability to put what economists refer to as "dead capital" to use. Dead capital, as the term suggests, comprises underused property. Consider a table saw, which costs about $300. For most households, this tool is probably used only a few times a year. In the past, people who did not want to spend such a sum for a one-off project could try to borrow a saw. This might have worked, but not everyone's friends or neighbors had a table saw or were willing to rent or lend it. This is no longer a problem because today, for a fraction of the cost, people can go on sites like NeighborGoods and rent saws and countless other tools from those who already own them. The sharing economy is so effective because it drastically expands a person's social network.

Beyond increased access, information is one of the hallmarks of the sharing economy. The rise of the Internet and the proliferation of smartphones exponentially increases consumers' access to information. This gives consumers a level of power that they have never had before.

Consider the following change in market dynamics. Each business transaction has three distinct parts: the buying decision, which is controlled by consumers; the selling decision, which is controlled by businesses; and information about the product or service. In the past, this information was controlled, or at least greatly influenced, by businesses. But, as author Jim Blasingame explains, all that has changed in today's economy because information is now in the hands of consumers.

Online feedback systems show how the costs of finding information have dropped dramatically. Web 2.0 – the generation of the web that introduced user interaction, sharing, and collaboration on sites – has led to a more consumer-friendly system. The sharing economy is just the natural extension of this, with its embrace of robust feedback systems.

Buyers frequently read reviews of products

to see whether other customers were happy or unhappy with their experiences. These reviews can be produced by trusted organizations and professionals, or they can come from the vast online community of past users. Peer-to-peer online interactions are like word-of-mouth reviews — only online interactions can reach exponentially more people than word of mouth ever could.

Even with all these benefits, the sharing economy is much more than a consumers' paradise. Workers also enjoy new advantages from the changing nature of work.

Americans have already absorbed many of the values of the sharing economy — particularly entrepreneurship. Indeed, millennials have been called the "start-up generation." A Bentley University survey finds that two-thirds of millennials have a desire to start their own business. And a Deloitte report shows that about seven in ten millennials envision working independently at some point in their careers.

Entrepreneurship has long been the primary escape from the tethers of traditional employment. But starting a business entails great costs and risks – such as quitting a steady job or building a business during the gaps of time between work, sleep, and other activities.

The sharing economy offers many of the benefits of entrepreneurship to a broader population, with lower cost and decreased risk. Independent contractors are the driving force behind the sharing economy's growth. Working as an independent contractor allows people to choose their own hours, and online platforms can provide work opportunities that fulfill the demand for independent work.

It is still hard to start a business and work for oneself. But, before the rise of the sharing economy, growing that business was even more difficult. This was especially true for niche products that needed to reach a customer base spread all over the world. By catering to producers of niche products, online platforms like the craft shop Etsy help to create widely successful independent companies. EBay has provided a similar type of benefit to sellers since 1995. In other words, the sharing economy enables greater levels of entrepreneurship because it makes it easier for millions of Americans to work for themselves.

For the last decade, the American economy has been generating more jobs in which workers are self-employed and contract out with other companies to offer their services. While people working for companies such as Uber and Airbnb still account for a small percentage of the US labor force, the individualized work arrangements that the sharing economy embraces make up a much larger, and growing, share of the labor force.

Research led by the American Action Forum's Will Rinehart finds that independent contractors accounted for nearly one-third of the jobs created from 2010 to 2014. Additionally, data from Christopher Koopman and Eli Dourado of George Mason University's Mercatus Center show that the number of 1099-MISC tax forms issued by the IRS to independent contractors increased by nearly

25 percent from 2000 to 2014, demonstrating that the shift toward independent work preceded the founding of Airbnb (2008) and Uber (2009). On the other hand, the number of W-2s, the tax forms used by salaried employees, slightly decreased over that same time.

Though the sharing economy is often

The key to understanding the sharing economy is that the real driving force behind the changes is not flashy smartphone apps but lower transaction costs.

---

associated with urban-dwelling millennials, older Americans also benefit from more flexibility and accessibility. Renting out homes and apartments with Airbnb and VRBO, hosting meals through EatWith, or driving for Uber or Lyft are all viable options for seniors to earn extra income. And the increased, on-demand access to goods and services also greatly helps those with limited mobility.

The key to understanding the sharing economy is the realization that the real driving force behind these changes is not flashy smartphone apps but lower transaction costs. Lower transaction costs affect every aspect of the economy.

Rather than adapting to these changes and embracing new technology, many established interests choose to fight economic advancement with politics. This tactic has been used relentlessly by hotels, taxi cartels, and labor unions against two of the most innovative parts of the sharing economy – home sharing and ride sharing.

Some Cities Can't Handle Innovation

Cities like Nashville, Tennessee, should love short-term home-rental platforms like Airbnb. For one, the city is a tourism magnet that remains short of enough hotels to meet surging demand. This is why Nashville's downtown has the highest average nightly hotel rate in the United States – ahead of cities like San Francisco, New York City, and Boston.

More importantly, some residents struggle to keep up with the increasing costs of living caused by steady economic growth. Unfortunately, rather than craft laws that embrace innovation and make it easier for both residents and tourists, Nashville's government has placed arbitrary limits on short-term rentals.

In 2015, Nashville bowed to pressure from special interests in the hotel industry and organized labor by passing an ordinance that restricted residents' ability to rent out their homes. The ordinance mandated a 3 percent cap on the number of nonowner-occupied homes in a neighborhood that could receive a short-term rental permit.

Under these new requirements, many long-time Airbnb hosts were unable to secure a permit. One of the families, P. J. and Rachel Anderson, could not get a permit because their neighborhood had already reached the 3 percent cap. The Andersons, who were long-time Nashville residents and Airbnb hosts, had the chance to relocate temporarily for a work opportunity in the music industry, something that is common in Nashville. However, they would be unable to afford to do so unless they could rent out their home while they were away. To fight back against Nashville's limits on renting nonowner-occupied homes, the family teamed up with the nonprofit Beacon Center of Tennessee and sued the city to regain their right to use their property.

In October of 2016, Circuit Court judge Kelvin Jones ruled in favor of the Andersons, striking down parts of the ordinance. While this was a positive result for all home owners, Jones's ruling did not stop restrictive laws from coming out of Nashville in the future. The ordinance limiting the Andersons from renting out their home was overruled for its unconstitutional level of vagueness, as it was unclear whether the Andersons' home qualified "as a hotel, bed-and-breakfast, or boardinghouse, which are not considered short-term rentals." However, the 3 percent cap was ruled to be constitutional and would be allowed to stay in effect if the ordinance were clarified and rewritten.

Critics allege that Nashville is being taken over by outside investors who buy houses only to rent them out through online platforms. However, an overwhelming majority of Airbnb hosts rent out the house that they live in. These types of property limits affect far more groups of people than simply wealthy outside investors. The Andersons' story is a telling example of how this ill-conceived attempt to crack down on what the city considers "illegal hotels" harms Nashville residents.

Another complaint against Airbnb in Nashville has been that it creates "party houses" and "illegally parked cars" in residential areas. While there are undoubtedly some disrespectful Airbnb guests, there are also countless rude full-time neighbors. The enforcement mechanisms that are currently used to hold residents accountable for actions such as holding late-night parties and playing loud music could easily be applied to short-term rental guests. Nuisance laws are already on the books — lawmakers do not need to solve a problem that already has a solution.

Furthermore, Airbnb's presence in Nashville helps to address a serious problem. Nashville currently hands out millions of taxpayer dollars to incentivize hotel building due to its hotel shortage. In this climate, Nashville policy makers would be wise to embrace the opportunity Airbnb and other short-term rental platforms offer for providing accommodations for visitors to the city. Yet, to satisfy the demands of special interests, they instead limit this technology's potential. And things may get worse, as the city has consistently considered placing a complete moratorium on issuing short-term rental permits.

Even if Nashville policy makers do not improve their short-term rental policy, the state of Tennessee can solve the problem. In the 2017 legislative session, a bill was introduced to create a statewide regulatory system that resembles the sensible regulations passed late last year by the Memphis City

Council. This preemption bill would prohibit local government from regulating homes that offer short-term rentals differently than homes that do not. This makes sense because if a local housing regulation is deemed necessary to protect the public or minimize nuisances, then it should apply to all housing – regardless of whether the home is rented out or not.

Arizona passed a law to this effect, SB 1350, that stops local governments from standing in the way of innovation when it comes to short-term rentals. Prior to the law's passage in Arizona, the areas of Jerome, Sedona, and Scottsdale had either banned short-term rentals or severely restricted them.

Florida is considering parallel legislation that would prohibit local governments from banning or restricting short-term rentals based solely on their "classification, use, or occupancy." This would be a win for Floridians, considering that there are currently dozens of local ordinances regulating short-term rentals. These regulations include fines of $10,000 a day, government inspections with only one hour's notice, exorbitant licensing fees, special utility and water assessments, excessively restrictive time frames for the use of private pools, and requirements for privacy and noise-buffering fences. As Florida senator Greg Steube, the sponsor of the state's preemption bill, says, "Local governments' goal is often to so heavily regulate short-term rentals that they are essentially prohibited."

Given what is happening across the country, it is difficult to argue with Steube's assessment. To show how these restrictions harmed Arizonans before the passage of the bill, Christina Sandefur and Timothy Sandefur of the Goldwater Institute tell the story of Glenn Odegard, an Arizona resident who bought a hundred-year-old house in Jerome, a struggling mountain town. As they write:

Glenn tried to contribute to that restoration by resuscitating a home that had been abandoned and left vacant for 60 years after a landslide filled it with rocks and mud. Intending to offer it as a vacation rental, Glenn lovingly restored the dilapidated house to its original historic condition. His successful efforts earned the home a feature in Arizona Highways *Magazine and a spot on the Jerome Historic Home and Building Tour. Yet despite issuing the relevant permits and initially embracing Glenn's home renovation, town officials decreed he could no longer use the home as a vacation rental. Under the town's newly announced ban, Glenn and other home owners face fines of $300 and up to 90 days in jail for each day they allow paying guests to stay. His "reward" for the investment of his time, money, and labor was to be considered an outlaw.*

A bill like Arizona's is likely necessary in Tennessee since the state's problems extend far beyond Nashville – Chattanooga inexplicably bans all short-term rentals in residential areas.

Nashville's history with short-term rentals makes it clear that state lawmakers should take this decision out of the City Council's hands. There is no reason for policy makers to be afraid to rein in cities that limit the economic opportunity that comes from technological progress.

Tennessee is not the only state that needs to protect property owners' rights. Just consider the conclusion that "there is no one-size-fits-all answer for municipalities in regulating short-term rentals" in the Colorado Municipal League's November 2016 report on short-term rental regulations. While this may be true, cities across Colorado show that there are countless misguided approaches to regulating the growing home-sharing industry.

On January 1, 2017, new regulations meant to limit short-term rentals in Denver went

---

There is no reason for policy makers to be afraid to rein in cities that limit the economic opportunity that comes from technological progress.

---

into effect. These restrictions, pushed by the hotel industry, represent unnecessary impediments to residents renting out their homes for some extra money.

The most problematic of Denver's rules is one that states that home owners are only allowed to rent out their primary residences. Even though most Airbnb hosts do rent out their primary residences, second homes are nonetheless an important part of the short-term rental industry. Second homes can comprise everything from vacation homes to long-time family homes that owners do not want to sell when work opportunities lead them to another city for a few years.

The second-most-active city in Colorado for Airbnb is Boulder, and Boulder County at large has certainly seen a major increase in growth as a result of Airbnb. Six of seven communities in Boulder County about doubled their number of annual Airbnb guests from 2015 to 2016. This is great news for Boulder County. Airbnb guests tend to stay more than twice as long and spend almost twice as much money in the community as other visitors. Airbnb's wide footprint in diverse neighborhoods also distributes travel income outside of traditional hotel districts.

One would expect cities to embrace this direct infusion of tourist dollars into the local economy, but Boulder has instituted similar regulations to Denver. The city requires hosts to rent out only their primary residence, and pay a hefty licensing fee and business taxes for doing so. Boulder also sets an occupancy limit of either three or four "unrelated persons."

Regulators are often at their worst when they get creative, and Durango's short-term rental restrictions are an example of this. The city's self-described "innovative policy solution" that "seek[s] to maintain neighborhood character, vitality, and vibrancy" is arguably the worst set of home-sharing regulations in the state.

Durango imposes strict restrictions on the

terms by which people can participate in this sector of the sharing economy. The city only allows short-term rentals in limited areas, and then caps the number that are available per block at one. In other words, it only takes one neighbor already having a short-term rental permit for Durango to not allow another family to rent out their home. All in all, Durango offers only sixty permits for short-term rentals for a town with eight thousand housing units. This nonsensical cap effectively shuts Airbnb and short-term rentals in general out of Durango.

In a state with such a vibrant and important tourism industry, Colorado's cities are constantly thinking of new "innovative" ways to partner with special industries. Instead of competing to see which city has the worst short-term rental policy, cities should welcome an industry that provides extra income to home owners, reduces prices for consumers, and boosts the economy of the region.

Colorado and Tennessee are not alone. Across the country, policy makers in everything from small tourist towns to America's largest cities have placed severe restrictions on short-term rentals. Opponents of home sharing focus on the local level, where city council members are more likely to fall for their flawed arguments. There is no denying that opponents have had high levels of success. If the cities continue to refuse to accept innovation, then state governments have a responsibility to follow Arizona's lead by stepping in and fighting back against special interests.

"The Rent Is Too Damn High"

In October 2016, New York passed a bill that creates civil penalties of up to $7,500 for advertising a whole apartment that is for rent for less than thirty days. This means that people going on a weekend getaway or staying with their significant other across town face thousands of dollars in fines for listing their place on Airbnb.

The justification for this bill was that Airbnb and other online short-term rental platforms drive up rents in New York City. This claim would be laughable if policy makers did not take it seriously – it is now the main argument used against short-term rentals.

First, there is no way that an online platform that did not launch until 2008 can be blamed for the city's decades-long struggle with high rents. How exactly is it possible that Airbnb could have been squeezing housing options for New Yorkers before its creation? The short answer is that it was not. In a classic case of government dishonesty, politicians blame problems created by years of government overreach on the new guy in town.

In an attempt to validate their war against short-term rentals, city leaders portray home sharing as a net loss of forty-one thousand housing opportunities (the number of active Airbnb nightly listings in the city) for locals. They claim that wealthy Airbnb users set up so-called illegal hotels and turn entire apartment complexes into unregulated temporary residences. This paints a dark picture of home sharing. After all, what could possibly be a better narrative for proudly progressive mayor

Bill de Blasio than the rich cutting corners at the expense of the poor?

The numbers quickly make this narrative fall apart. Airbnb's forty-one thousand active nightly listings in New York City come out to just over 1 percent of the city's three million housing units. Additionally, in New York City, 90 percent of Airbnb posts are for residents' permanent homes. There is no way that taking a maximum of 0.1 percent of New York City's residential units off the market by using them exclusively for short-term rentals is what drives rent increases. This is especially true considering that the city has 110,000 hotel rooms. These numbers should make people question the sudden concerns over access to affordable housing from prominent leaders in the hotel industry.

The real reason for higher rents is a combination of a lot of people wanting to live in New York and an insufficient amount of housing supply to meet this demand. High levels of regulations that limit development are a major contributor to the lack of housing

> *Land use regulations are now so overreaching that 40 percent of the existing buildings in Manhattan would not be able to be built today.*

supply.

Land use regulations prevent denser construction and building more units in existing buildings. Construction and zoning regulations have drastically increased in scale and scope. They are now so overreaching that 40 percent of the existing buildings in Manhattan would not be able to be built today. With a high level of restrictions on how builders can meet the demand for housing, it is no wonder that New York City property is so expensive.

Simply put, living in Manhattan will always be more expensive than living in rural

Oklahoma. Furthermore, there is no way to combat the housing affordability crisis without expanding the supply of housing. It is not Airbnb that causes high rents but politicians.

If anything, Airbnb is an asset to the middle class. The average Airbnb host in New York City makes about $5,500 a year from the service, money that 76 percent of users say helps them stay in their homes or apartments.

The claim that Airbnb is to blame for high rents is also common across cities in California. Here again, housing shortages in San Francisco and Los Angeles were instead caused and perpetuated by city governments' unwillingness to increase the housing supply.

Since the 1960s, the residential construction rate in California has significantly declined and, as a result, real housing prices increased by 385 percent from 1970 to 2010. Though limits on new construction are not the sole cause for increasing housing prices, they are a major factor.

The least-affordable housing markets are those where new housing permits have not kept up with population growth. For example, in Los Angeles, there were twenty-eight thousand housing projects started in 2014. Meanwhile, Houston, a city with no housing shortage and 1.7 million fewer people than Los Angeles, started sixty-four thousand housing projects in 2014. The lack of housing in Los Angeles is not due to overcrowding – the city has roughly eight thousand people per square mile, which is one-third of New York City's level.

In Los Angeles, home owners spend an average of 40 percent of their income on mortgage payments, while San Francisco residents spend an average of 48 percent on them. Airbnb provides an opportunity for these home owners to meet their mortgage obligations by helping them rent out spare rooms.

San Francisco, where rents for a one-bedroom apartment frequently exceed $4,000 per month, has the most serious housing shortage in America. Over the past twenty years, San Francisco has only permitted the annual construction of an average of 1,500 housing units. Over that time, San Francisco's population grew by ninety-seven thousand, and growth has been stronger in recent years.

A study by the real estate firm Trulia finds that San Francisco had the highest median rental prices per square foot and the lowest rate of new construction permits among America's ten largest tech hubs. A reason for this is that nearly 80 percent of San Francisco's housing is occupied by rent-controlled tenants or home owners. This leaves only one in five housing units available for other renters, artificially driving up rents.

Additionally, the booming, high-salary tech industry represents about 8 percent of the workforce in San Francisco, putting further upward pressure on the price of housing in an already-overburdened market. This is why some people blame tech workers for the city's housing shortage. One proposed solution is for tech companies such as Google to create more housing for employees on company property. Although this seems to be logical, the city of San Francisco explicitly forbids it.

San Francisco, unlike many other major US cities, has building permits that are discretionary rather than as-of-right. This standard makes it more difficult to gain approval for development because of the numerous legal challenges that it invites. For new housing developments in San Francisco, there is a preliminary review, which takes six months. Then there is a chance that neighbors will appeal the permit on either entitlement or environmental bases. These barriers add unpredictable costs and years of delays for developers, the financial responsibility for which is ultimately passed on to buyers and renters.

It is easy for politicians to blame corporations for problems that are ultimately their own responsibility. This is precisely what California senator Dianne Feinstein has done. In October 2015, she published an article in the *San Francisco Chronicle* arguing in favor of the city's Proposition F, which would have limited short-term rentals in San Francisco to seventy-five days a year.

According to Feinstein, this ballot initiative would have helped to alleviate the city's housing shortage. Predictably, Feinstein did not mention the lack of housing permits in cities across her state. It is also important to note that she and her husband have a stake in a San Francisco hotel that is worth up to $25 million. Proposition F was soundly defeated at the ballot box, but city leaders quickly moved to create other restrictions on short-term rentals.

Ironically, Airbnb might owe its existence to San Francisco's housing shortage. In 2007, the city did not have enough hotel capacity to house visitors for an industrial design conference, so Airbnb's founders Brian Chesky and Joe Gebbia decided to start a company to fix this problem.

Another tactic taken by hotel unions, such as the AFL-CIO–affiliated New York Hotel and Motel Trade Council, is to claim that Airbnb threatens "good-paying union hotel jobs in New York City and around the country." Despite the union's claim that hotels across the nation will suffer, the data indicates the overall hotel industry is managing quite fine even with the growth of home sharing. According to STR, a leading company in hotel market research, the hotel industry just had its best year on record. Hotels and home sharing can thrive together because companies like Airbnb expand the proverbial pie of lodging options. Both models offer different experiences, levels of convenience, and price points.

Airbnb helps to relieve some of the symptoms of high rents and housing shortages by giving people some help with their existing rents or mortgages. The claim that Airbnb is the reason for higher costs of housing is demonstrably false. Instead of placing blame on Airbnb for housing shortages, politicians should embrace the service and evaluate their regressive restrictions on new housing development.

* * *

A Tech Capital without Uber

Austin is home to the University of Texas at Austin, countless start-ups, and the widely popular South by Southwest festival. But even though the city is commonly referred to as the tech capital of the South and has a reputation as a millennial paradise, Uber and Lyft ceased operations there in May of 2016. Austin is the largest city in the United States without Uber or Lyft.

Local policy makers blame Uber and Lyft for leaving the city, but the Austin City Council pushed out ride sharing by regulating in search of a problem. Austin's case of regulatory overreach shows how even the country's most progressive cities are often hostile to economic progress.

In December of 2015, the Austin City Council approved an ordinance to require ride-sharing drivers to go through fingerprint background checks. Rather than altering their business models, Uber and Lyft took their protests to the voters by setting up a ballot challenge to the ordinance called Proposition 1. The proposition failed, and the companies ended their service in the city.

Complying with fingerprinting requirements would not have increased public safety. But doing so would have made it more difficult for ride-sharing companies to quickly and equitably provide work opportunities as they grew to meet ever-increasing consumer demand. Even though fingerprint background checks may sound like they increase security, they are unnecessary, ineffective, and discriminatory when used for job screening instead of for law enforcement.

Uber and Lyft both use name-based background checks instead of government fingerprinting. The companies' safety record shows that this approach works. Indeed, a comprehensive review of these name-based background checks by the Maryland Public Service Commission finds that the companies' checks are just as "comprehensive and accurate" as government fingerprint background checks. This review came after the state passed a law in 2015 that would have required fingerprinting for all ride-sharing drivers if the commission reached the opposite conclusion.

Instead of placing blame on Airbnb for housing shortages, politicians should embrace the service and evaluate their regressive restrictions on new housing development.

Policy makers who insist on fingerprint background checks must be watching too much *CSI*. These checks are far from the foolproof tool that they are portrayed as on TV. As the Cato Institute's Matthew Feeney argues, "Fingerprinting is not as effective as Hollywood might have you believe. The FBI fingerprint database is incomplete, in part because it relies on police departments and other local sources adding relevant data and keeping that data updated."

In other words, fingerprint background checks are only as effective as the databases of fingerprints that they pull from. In Maryland, certain traffic violations – including DUIs and reckless-driving incidents – would not show up through fingerprint background checks. Name-based background checks avoid this failure by querying thousands of courthouse and law enforcement databases to find relevant records.

Fingerprint databases also include arrests that did not lead to any convictions. If someone is arrested and fingerprinted but then found not guilty or never charged with a crime, that information would need to be updated by law enforcement for fingerprint background checks to be effective. Yet this follow-up step is often overlooked, which leads to discriminatory results.

Law enforcement personnel do not have the time to keep these records updated because fingerprint databases were designed to aid in solving crime – not vetting for-hire vehicle drivers. Since about one in three felony arrests end up with no conviction, fingerprinting leads to many qualified drivers without criminal convictions being denied work opportunities. The Urban League, NAACP, and National Black Caucus of State Legislators all oppose fingerprinting requirements for this reason.

Fixing these errors is difficult. University of Maryland professor Michael Pinard testified at a public hearing about the requirements that it would be "difficult if not impossible" for people without law degrees to challenge the faulty results from a fingerprint background check.

The most troubling result of Austin's ride-sharing regulations is not that two successful start-up companies will lose revenue by pausing their operations in the city – it is the new problems and dangers posed to Austin residents.

No background checks are perfect, but ride sharing's safety record shows that the companies' current policies are working well. The safety standards that Uber and Lyft voluntarily hold themselves to are even stricter than those required of Austin taxis. In addition to providing safer transportation than taxis do, ride sharing's business model makes it very easy to police drivers. The locations of both parties are tracked through the duration of the trips, identities are verified, and no cash changes hands because of the platform's electronic payment systems. In other words, if drivers commit crimes while working for Uber or Lyft, they must be doing so because they want to get caught.

When evaluating the potential risks of new technologies, it is important to weigh them against the real dangers posed by the status quo. For example, New York City's police commissioner, Bill Bratton, told women to use the "buddy system" when riding in the city's taxis. This is because taxi drivers are a leading perpetrator of sexual assaults by strangers, with fourteen women being raped by New York City taxi drivers in 2015. There will undoubtedly be isolated cases of bad behavior by ride-sharing drivers, but the built-in safeguards make it much easier to hold bad actors accountable, which discourages dangerous conduct.

Rather than increasing public safety by mandating fingerprint background checks, Austin policy makers are placing their constituents at greater risk. Perhaps nowhere is this unintended consequence clearer than with drunk driving. Ride sharing has been documented to lower both drunk-driving arrests and fatal accidents, partly because taxis are difficult to find late at night. In Austin – a city with the highest number of downtown bars per capita in the United States – the number of available taxis drops at midnight. This is when alcohol-related crashes and DUI arrests are at their highest levels. It should therefore come as no surprise that Mothers Against Drunk Driving is a major proponent of ride sharing as an additional reliable source of transportation.

Furthermore, as Austin resident and Texas

Public Policy Foundation senior fellow John Daniel Davidson has chronicled, getting around the city has become much more challenging. During University of Texas football games, the city's many music festivals, or just regular Saturday nights, the streets are chaotic. Even though there are new ride-sharing services that require fingerprinting for their drivers, these applications are often unreliable and getting a ride is very difficult during times of high demand. And, of course, public transit options in the city remain terribly inadequate.

In addition to unfounded public safety concerns, one of Austin's main justifications for fingerprinting Uber and Lyft drivers was that the city's taxi drivers were required to go through the same process. The new regulations were nothing more than a way to harmonize regulations between all forms of for-hire vehicles, according to proponents of further regulation.

Massachusetts senator Elizabeth Warren made such a claim in a speech at the New

America Foundation, asserting that Uber is "[fighting] against local rules designed to create a level playing field between themselves and their taxi competitors."

But there are two ways to level the playing field. One is to apply antiquated regulatory requirements to new technologies, as Austin did. The far-superior option is to embrace innovation by getting rid of pointless requirements that tend to protect established businesses rather than consumers, such as taxi permits that limit the supply of available cabs or ineffective and discriminatory fingerprinting requirements.

Taxis in Austin are still required to get a government permit to operate. But the city, which has a population of nearly nine hundred thousand people, only issues 915 taxi permits in total. Furthermore, these permits are owned by just a handful of companies.

Until Uber and Lyft came along, all these barriers to entry led to a lack of competitive pressures – and customer service in the taxi industry noticeably suffered. Competition allows companies to differentiate themselves through the type, price, and quality of services they offer. Rigid regulations that dictate the means transportation companies can use to meet customers' needs create a one-size-fits-all standard.

To help taxis better compete against their ride-sharing competitors, local policy makers should focus on cutting the red tape so that taxis can become more like Ubers – not the other way around. As Michael Farren of the Mercatus Center argues, if policy makers must apply regulation to new business models such as ride sharing, a better way forward would be to specify certain areas of concern (such as prescreening for quality and safety, or clarity in pricing) but then leave companies free to figure out how to meet these requirements.

Texas is not the only state where the fingerprint background check debate has left people without ride sharing. Cities in upstate New York are facing the same fight.

This is surprising given what Governor

Andrew Cuomo had to say about ride sharing. Back in July 2015, the governor asserted, "Uber is one of these great inventions, start-ups, of this new economy and it's taking off like fire to dry grass and it's giving people jobs. I don't think the government should be in the business of trying to restrict job growth."

Two years later, New York's government still does not allow ride-sharing services to operate in upstate New York. This inability to adapt to new business models leaves millions of New Yorkers with fewer transportation options and work opportunities.

Besides Austin, Buffalo is the largest US city without Uber or Lyft. Buffalo's poverty rate of 31 percent is nearly two-and-a-half times greater than the average rate in the United States. The city's five-year labor force participation rate, which measures people who are working or looking for work, stands at 59 percent, far below the average US rate. These are all problems that could be addressed with more work opportunities, which Uber and Lyft would provide.

And Rochester, another major city in New York that does not have ride sharing, has likewise paid the price for this choice economically. The city, though it is known for its colleges and universities, has one-third of its residents living in poverty.

Just how much economic activity are these New York cities missing out on by not having ride sharing? In November 2015, Uber estimated that the company would create an additional thirteen thousand jobs if the company expanded to all of New York. If anything, this job-creation estimate is too low;

The best way to level the regulatory playing field is by getting rid of pointless requirements that tend to protect established businesses rather than consumers.

Uber has thirty thousand active drivers in Pennsylvania, a state with a similar population to that of New York when excluding New York City, the only area of the state where Uber and Lyft operate.

A February 2017 study by Land Econ Group finds that if Lyft started operating in Buffalo and Rochester in 2014, local drivers could have earned over $18 million from partnering with the company in 2017. The projection is based on the number of observed rides in cities with comparable populations, incomes, and densities. This is real income that would have helped those struggling to make ends meet in the sluggish upstate New York economy.

It is true that most new ride-sharing jobs would be part time. Half of Uber drivers work on the platform for under ten hours a week and 80 percent of Lyft drivers work on the platform for under twenty hours. Some critics of ride sharing's business model worry about part-time work, but flexible work is a feature of ride sharing – not a bug.

Because drivers set their own schedules and use their own cars, a wide array of people benefit. From single parents and full-time drivers to college students and retirees, virtually anyone can earn additional income by partnering with ride-sharing companies. This is why nearly all Uber and Lyft drivers cite flexibility as the main benefit of contracting with the companies.

New Yorkers realize these diverse benefits even if some of their elected representatives do not. A 2016 Siena College poll finds that "at least two-thirds of voters from every [New York] region and every [political] party support legislation to allow ride-sharing companies such as Uber to operate in their areas." Overall, support for legalized ride sharing was at 80 percent at the time of the poll.

In 2017, there is no excuse for major cities to leave their residents without the work opportunities and transportation options that ride sharing provides. At least thirty-seven states currently have statewide regulatory frameworks that allow ride-sharing services. While some of these frameworks are far from ideal, at least they allow ride sharing's business model to continue to operate and develop. Of course, New York and Texas are two of the states that do not have such a framework.

As the experiences of hundreds of other cities throughout the world show, ride sharing is a safe and affordable addition to a city's transportation infrastructure. Yet, some big-city mayors see the service as an affront to their progressive values. But how can companies that provide thousands of residents with jobs, increase access to transportation options (especially in underserved areas), and lower prices possibly be anything but progressive? Where the regressive policy option is to regulate ride sharing out of existence, the progressive choice is for city leaders to resist the urge to overregulate and instead embrace services that improve the lives of their residents.

* * *

Unions Can't Stand the Sharing Economy

Seattle may have allowed unions to negotiate away an entire profession when, in December 2015, the Seattle City Council voted to extend collective-bargaining rights to certain independent contractors, including those who partner with ride-sharing platforms such as Uber and Lyft. This move was the first of its kind, because federal labor law only allows employees to collectively bargain, which means having an organization negotiate pay structures and working conditions for an entire class of workers.

If successful, Seattle's collective-bargaining process will silence the opinions of drivers who may not have chosen to be in a union but will be forced to join one anyway. And once Seattle's model spreads to other cities, collective bargaining will mean many of the sharing economy's benefits will cease to exist.

The outdated union model is antithetical to the flexible, entrepreneurial workplace that many Americans – especially millennials – desire. Ride sharing is a popular work opportunity for a multitude of reasons that collective bargaining will likely limit. If the vote is successful, all drivers who want to keep working will be forced to join a union and follow the collectively bargained agreement.

Under union representation, new workers are often the first to be fired, even when they perform better than those who have more experience. Would a union defend bad drivers who had their Internet accounts deactivated because of negative reviews? Post-ride, dual-feedback systems are major factors behind ride sharing's increased levels of customer service and trust. If partnerships with unqualified drivers cannot be terminated, riders will be less safe and the incentive for drivers to offer a pleasant riding experience will be reduced.

The other union norm of maximum or minimum work hours would take away drivers' freedom to work as much or as little as they desire. For example, some drivers prefer to drive early in the morning on weekdays to avoid traffic and get lucrative airport trips. Others work during bar closing times to take advantage of the higher fares that come with the increased demand. The scheduling put in place by unions would likely favor full-time, established drivers and take away the diverse benefits that the ride-sharing model provides. Predetermined driving schedules that cannot adjust to demand would also lead to drivers idling, unable to find fares, and riders waiting on the side of the road, unable to find available cars.

There is no indication of what the costs of belonging to a union will be if collective bargaining is successful. But there is reason to worry that some part-time or seasonal drivers will lose money during the time periods that they do not drive because they still must pay the union. For example, some college students and teachers only drive during summer vacation. There is no reason that they should have to pay dues and lose money when they are not actively driving.

There is also no guarantee that the initiation fees and dues will be the same for all for-hire vehicle companies. One could imagine a situation where it costs much more to drive with Lyft than with Uber or a taxi company. This imbalance could give a lot of power to the companies with the lowest union fees, which could lead to fewer options for both riders and drivers.

Drivers are very sensitive to increases in barriers to starting work. One of the main benefits of the ride-sharing model is that there are low start-up costs. Every added cost or delay, whether fingerprint background checks or union initiation fees, makes it less likely that people will begin partnering with ride-sharing companies. The effects will be even worse for drivers who utilize multiple platforms and thus would have to pay multiple sets of dues. This practice is common, as more than 40 percent of Lyft drivers also drive with other platforms.

The National Labor Relations Act, passed in 1935, established the right to collectively bargain for all employees. This right does not apply to the independent contractors who make up the sharing economy. Independent contractors are free to join groups such as the Freelancers Union or the App-Based Drivers Association to gain access to benefits and career-development resources, but they cannot collectively bargain.

Excluding independent contractors from collective bargaining makes sense. Even people who only work with one company do not all have the same priorities regarding work arrangements. Those who use Uber for supplemental income and part-time work have vastly different concerns than those who treat the service as a full-time job. Under collective bargaining, which group's interests will the union represent? Majority rule could take away one of the cornerstones of the sharing economy: the diverse benefits that come from flexible, individualized work opportunities.

The total flexibility is what makes working in the sharing economy so desirable in the first place. The ability to quickly increase earnings to meet rent, pay down student loans, or fund a new business venture benefits people every day. This "income smoothing" is especially critical for young people and the poor, groups whose earnings often fluctuate wildly. Some 70 percent of Americans between the ages of eighteen and twenty-four, and 74 percent of those in the bottom income quintile, experience an average of 30 percent or more in income changes from month to month, according to the JPMorgan Chase Institute. Ride sharing's flexible work model empowers people with diverse priorities to help themselves and those counting on them.

Troublingly, because Seattle's rules require unions to negotiate in areas such as drivers' earnings, terminations, and hours, collective bargaining in the city could eventually lead to courts deeming drivers to be employees instead of independent contractors. The more control unions force the companies to have over drivers, the more likely this scenario becomes.

This is not what drivers want. An independent survey of 3,100 Lyft drivers finds that 82 percent of respondents agreed or strongly agreed with the statement "I like being an independent contractor." Since flexibility is one of the main benefits of the ride-sharing model, it is not surprising that 99 percent of Lyft drivers agreed that with the statement "I like to choose when I work."

Uber drivers share the same sentiment. When six hundred Uber drivers were asked the question "If both were available to you, at this point in your life, would you rather have a steady 9-to-5 job with some benefits and a set salary or a job where you choose your own schedule and be your own boss?" 73 percent said that they prefer flexibility over the traditional employment model. Opponents in organized labor fail to realize that the flexibility of being an independent contractor is vital to the sharing economy's success.

Union opposition to the sharing economy is widespread, and it extends beyond Uber and Lyft. The International Brotherhood of Teamsters explains its perception of the effects

> *Under collective bargaining, majority rule could take away one of the cornerstones of the sharing economy: the diverse benefits that come from flexible, individualized work opportunities.*

of the sharing economy in its newsletter, writing, "[Sharing-economy] companies are simply recycling old ideas and taking us backwards to a time when workers had no rights on the job." The Teamsters' complaints do not stop there, as the newsletter also warns, "Don't let the term 'sharing economy' fool you. There is no sharing. It's really just the one percent making money by stripping workers of the rights for which the labor movement has fought so hard to secure."

It is unsurprising that union advocates lament the sharing economy on the basis of reducing union membership. Naturally, union advocates seek to prevent work that falls outside the realm of a typical union job. The 1950s steel worker had his job protected and enhanced by a union, so the 2017 Uber or Lyft driver must supposedly conform to union association as well, according to their arguments.

In February 2016, the AFL-CIO, America's largest labor union, released a statement that made clear how its leadership views sharing-economy workers — as a potential boon to union member rolls. "Making the right policy choices begins with ensuring people who work for on-demand companies enjoy the rights and protections of employees," the statement reads. "The AFL-CIO is committed to ensuring new technology — and new forms of employer manipulation — do not erode the rights of working people. Rest assured that if employers get away with pretending their workers aren't employees, your job could be next."

Unions are desperate because worker

preference and demographics are working against them. Today, only 11 percent of American workers (6 percent in the private sector) belong to unions. Unions have exerted power beyond their small numbers due to their outsize campaign contributions, practically all to Democratic politicians. To keep the flow of contributions and their power, and to prop up their underfunded pension plans, unions need more dues-paying members. Yet fewer Americans, particularly younger workers, are joining unions.

Labor Department data show that 14 percent of workers between the ages of forty-five and sixty-four belong to unions, compared with 10 percent of twenty-five- to thirty-four-year-old workers, and a historically low 4 percent of those aged sixteen to twenty-four. Most millennials may be prounion in theory, but when it comes to parting with a portion of their paychecks, few are willing to support unions in practice.

Labor unions have lost sight of the reality that to remain viable they must meet young workers' needs. Instead, their focus is to satisfy the demands of union bosses and retirees, as evidenced by bloated union-leader salaries and unsustainable pension promises.

Of course, to unions, it does not matter whether workers even desire collective-bargaining agreements that erode their independence and flexibility. This type of opposition to independent work on the part of unions is epitomized by Robert Reich, who has called the sharing economy a "fraud" that "should be called the 'share-the-scraps economy.'" Reich, professor of public policy at University of California Berkeley and former secretary of labor under President Clinton, maintains that the sharing economy is "a reversion to the piece work of the nineteenth century – when workers had no power and no legal rights, took all the risks, and worked all hours for almost nothing."

But who is Reich to decide that the millions of Americans who have voluntarily partnered with a sharing-economy company are being ripped off? Though he is not an economist by training, he should be familiar with the economic concept of "revealed preference." People choose to partner with sharing-economy companies because, in their view, doing so is the most desirable option available to them. To make this more difficult for them to do is to essentially say, "I do not care if you enjoy driving with Uber as an independent contractor; I disapprove of your decision and will not allow it."

Instead of obsessing over trying to revive the post–World War II era of high unionization rates, politicians should embrace the emerging flexible workplace. Besides the benefits that the sharing economy offers to workers, consumers, and the economy, it is in lawmakers' political interest not to intervene in the new economy's ascent.

The Politics of the Sharing Economy

Back in 2013, at the Eighty-First United States Conference of Mayors, a bipartisan group of nine mayors from major US cities released a resolution urging "support for making cities more shareable."

Additionally, a June 2015 National League of Cities survey of 245 elected officials found that over 7 in 10 were supportive of the sharing economy, and not just because of the work opportunities it creates. The most popular response when leaders were asked to name the biggest benefit was "improved services."

Despite this past optimism and positive rhetoric, excessive regulatory barriers or outright bans still plague the sharing economy in certain cities because small, concentrated interests use the political process to receive special treatment. This problem is especially pronounced on the local level, where hotel owners, taxi cartels, and labor unions have their highest levels of influence.

In the long term, opposition to Uber, Airbnb, and the rest of the sharing economy is a losing political strategy, as Americans for Tax Reform's John Kartch argues. The Pew Research Center finds that potential young voters (ages eighteen to twenty-nine) are nearly five times more likely to use ride sharing than Americans who are at least fifty years old. Pew's research also shows that Americans of all ages and political leanings overwhelmingly reject the idea of applying the same regulations that hamper legacy industries to new services like ride sharing.

Furthermore, a Reason Foundation poll finds that only 18 percent of millennials have faith in government regulators to have the public's best interests in mind. It takes a lot to get young people to vote, but if policy makers get between millennials and their Ubers or Airbnbs, there will be major consequences at the polls in the future. As these services continue to expand, the political costs of opposing the sharing economy will only grow.

In financial and energy industries, regulations' negative consequences are far from clear to consumers. For example, most people do not know how much specific EPA regulations will increase the cost of charging a smartphone. Similarly, the effects of Dodd-Frank's financial regulations are far from clear when someone sends money on Venmo or applies for a mortgage through an app. But with Uber and Airbnb, the negative effects of government intervention are crystal clear — higher prices, fewer options, and reduced earnings opportunities.

Young Americans rightly realize that many regulations do little more than protect established interests. Furthermore, following in the footsteps of their parents and grandpar-

---

Young Americans realize that many regulations do little more than protect established interests. Millennials appreciate the value of entrepreneurship and most hope to work for themselves in the future — a goal that the sharing economy can help realize.

---

ents, millennials appreciate the value of entrepreneurship. The majority of young Americans hope to work for themselves in the future – a goal that the sharing economy can help realize.

Though fights over Uber and Airbnb regulations often make the news, it is important to realize that the same trends that enabled these companies' business models will transform more than vehicle transportation and short-term lodging. Technology is going to continue making the workforce and marketplace more flexible, individualized, and mobile. Americans are entrepreneurial by nature, and innovations that make it easier for people to work for themselves while providing more options to consumers should be welcomed by policy makers.

The essence of the sharing economy is consumer empowerment, and increased connectivity has transformed the relationship between consumers and service providers for the better. The world we live in today is one that prizes accessibility and convenience.

It is a world that has taken power from large corporations and multinational suppliers and given it to consumers through the widespread availability of information.

Yet, in the face of these benefits, fearful regulators in cities across America continually claim that consumers are hurt by the trust they have rightly placed in the new business models that characterize the sharing economy. Rather than keeping consumers safe, local regulators now threaten the growth of the peer-to-peer system that has proven to be the most effective way to increase consumers' access to information.

As the economic trends that drive the sharing economy expand their reach into more and more industries, regulators will surely continue to find themselves in uncharted territory, and their decisions will have major effects on the level of economic growth. When these regulators encounter new business models, they need to keep in mind that the technological progress behind the sharing economy is a positive development. The sharing economy is driven by lower transaction costs, and waging a war on lower transaction costs is the definition of fighting progress.

If mayors and city councils across the United States continue to oppose the sharing economy, state policy makers have a responsibility to stand up for innovation. This may require overruling cities when local policies threaten the development of the new economy. Truly progressive thinking means not standing in the way of clear innovation and widely shared growth. If cities are going to remain a driving force for economic progress, then states need to save cities from themselves.

© 2017 by Jared Meyer

All rights reserved. No part of this publication may be reproduced, stored in a retrieval system, or transmitted, in any form or by any means, electronic, mechanical, photocopying, recording, or otherwise, without the prior written permission of Encounter Books, 900 Broadway, Suite 601, New York, New York, 10003.

First American edition published in 2017 by Encounter Books, an activity of Encounter for Culture and Education, Inc., a nonprofit, tax exempt corporation.
Encounter Books website address: www.encounterbooks.com

Manufactured in the United States and printed on acid-free paper. The paper used in this publication meets the minimum requirements of ANSI/NISO z39.48–1992 (R 1997) (*Permanence of Paper*).

FIRST AMERICAN EDITION

LIBRARY OF CONGRESS
CATALOGING-IN-PUBLICATION DATA
IS AVAILABLE

SERIES DESIGN BY CARL W. SCARBROUGH